HOW MANY PENGUINS?

Jill Esbaum

Counting Animals

2 3 4 5 6 7 8 9 10 11 12 13 14 15 16 17 18 19 20 21 22 23 24 25

Children's Press®
An imprint of Scholastic Inc.

Library of Congress Cataloging-in-Publication Data
Names: Esbaum, Jill, author.
Title: How many penguins? : counting animals 0-100 / Jill Esbaum.
Description: New York : Children's Press, an imprint of Scholastic Inc., [2022]. | Series: Nature numbers | Includes index. | Audience: Ages 5-7. | Audience: Grades K-1. | Summary: "Nonfiction, full-color photos of animals and nature introduce basic math concepts and encourage kids to see a world of numbers all around them"– Provided by publisher.
Identifiers: LCCN 2021031695 (print) | LCCN 2021031696 (ebook) | ISBN 9781338765182 (library binding) | ISBN 9781338765199 (paperback) | ISBN 9781338765205 (ebk)
Subjects: LCSH: Counting–Juvenile literature. | Animals–Miscellanea–Juvenile literature. | BISAC: JUVENILE NONFICTION / General | JUVENILE NONFICTION / Mathematics / General Classification: LCC QA113 .E74 2022 (print) | LCC QA113 (ebook) | DDC 513.2/1–dc23
LC record available at https://lccn.loc.gov/2021031695
LC ebook record available at https://lccn.loc.gov/2021031696

Series produced by WonderLab Group, LLC
Book design by Moduza Design
Photo editing by Annette Kiesow
Educational consulting by Leigh Hamilton
Copyediting by Jane Sunderland
Proofreading by Molly Reid
Indexing by Connie Binder

Photos ©: cover: Vladimir Seliverstov/Dreamstime; 4-5: Vladimir Seliverstov/Dreamstime; 6-7: Yva Momatiuk and John Eastcott/Minden Pictures; 10-11: Karen Christensen/Dreamstime; 14-15: Ali Sulima/Dreamstime; 16-17: Tarpan/Dreamstime; 20-21: Javarman/Dreamstime; 23: Edwin Godinho/Dreamstime; 24: Cyril Ruoso/Nature Picture Library; 26-27: Sylvain Cordier/Nature Picture Library.

All other photos © Shutterstock.

10 9 8 7 6 5 4 3 2 1   22 23 24 25 26

Printed in the U.S.A.   113
First edition, 2022

For William, Bennett, and Hallie
—JE

Try It!

Look for answers to all the "Try It!" panels on page 31.

Imagine you are a fuzzy penguin chick.
When will you start to look like your sleek parents? Penguin chicks are born with downy feathers. Then they <u>molt</u> and grow new feathers. In a year, they will have the waterproof feathers of adult penguins.

Count the penguins. Did you count more or less than 10?

emperor penguins

Antarctica

Antarctic fur seals are most at home swimming and diving in the ocean. But they sometimes come ashore to rest.

Try It!

Count the Antarctic fur seals. If **6** of them slipped into the water, how many would be left?

Antarctic fur seals

South Georgia Island

Hi there, monkeys.

Hope you don't mind resting while we count you.

Go ahead and keep <u>grooming</u>. Picking dirt and bugs out of your fur helps keep you healthy.

8

langur monkeys

Indian forest

Try It!

How many monkeys? Count them up. From that number, keep counting up to 20.

Red rock crabs
are easy to see
but hard to catch.
Their ten legs help
them move quickly!

red rock crabs

Amazon Basin

Try It!

Count the crabs on this rock. If 11 more joined them, how many would there be?

Burrowing owls
can't see so well out of
the sides of their eyes.
So they <u>swivel</u> their heads
almost in a full circle to see.
These birds hunt on the ground
near their underground
nests. They look for insects
and rodents to eat.

Try It!

Count the owls. Pretend a
neighboring <u>burrow</u> has the same
number of owls. How many owls
would that be, all together?

burrowing owls

Colorado prairie,
United States

13

A week without water?
No big deal.

A month without food?
Camels can take it.

When a camel has lots to eat, its hump stores fat. A camel's body turns that fat into <u>energy</u> to keep it going strong when there is less food. When a camel goes a long time without eating, its hump actually <u>shrinks</u>!

camels

Arabian Desert

Try It!

Count the camels on the left page. Count the camels on the right page. If you subtract the smaller number from the larger one, how many camels are there?

reindeer

Russian tundra

In spring, reindeer herds roam Arctic lands, nibbling anything green and growing.

Winters are freezing cold here. Good thing these reindeer have thick layers of hair.

Their sensitive noses can smell food under the snow. Sniff-sniff, scrape-scrape-scrape . . .

Ta-da! A lichen lunch!

Try It!

Count the reindeer in this small herd. If 10 more reindeer come along, how many will there be?

puffin

Farne Islands,
United Kingdom

Puffins are hard workers. All day long, parents fly off and dive for fish. Then they bring them back to their hungry chicks.

How many fish can they carry at a time? **Ten or more!**

Try It!

Imagine a puffin parent flying off 3 times. Each time, it brings back 10 fish. How many fish would it bring back, all together?

A family of elephants is called a herd. Who's in charge? The oldest female. Because of her long life, she remembers things the others may not.

Try It!

Count the elephants. If this herd usually has 12 elephants, how many have wandered off?

She remembers: Where to find food and safe places to sleep. How to make calves behave. Which watering holes are perfect for a family drink.

African elephants

South African savanna

Giraffes huddle close together to be safe.

This makes it harder for a lion to pick one giraffe and hunt it down.

Try It!

How many giraffes are in this photo? Now add 14. Is the total number more, less, or the same as 20?

giraffes

Tanzanian savanna

ring-tailed lemurs

Madagascar

24

Ring-tailed lemurs always travel in groups.

If members stray too far apart, one calls them all back together with a **loud meow!**

Try It!

How many lemurs are there in this family troop? Count up from that number until you reach 50.

monarch butterflies

Mexican forest

Monarch butterflies <u>migrate</u> long distances in very large groups.

You've counted so many animals. How high can you count now?

Go from zero to 100!

53 54 55 56 57 58 59 60 61

62 63 64 65 66 67 68 69 70

71 72 73 74 75 76 77 78 79

80 81 82 83 84 85 86 87 88

89 90 91 92 93 94 95

96 97 98 99 **100!**

Try It! Activities

In this book, readers visited and counted animals living in a variety of habitats around the world. Here are counting activities kids can do to take the fun beyond the pages of this book.

EMPEROR PENGUINS (pages 4–5) walk by waddling. Count your steps as you walk normally across a room. Now count your steps as you waddle like a penguin across a room. Which kind of walking took more steps?

ANTARCTIC FUR SEALS (pages 6–7) sometimes make growling sounds. How many different animal sounds can you make?

MONKEYS (pages 8–9) hang from branches using their hands, feet, and even their tails. How many seconds can you hang by your hands from a strong bar? Have somebody count for you.

RED ROCK CRABS (pages 10–11) have ten legs. How many fingers do you have? How many toes? If you add them all up, how many do you have, all together?

BURROWING OWLS (pages 12–13) hurry into their burrows when they're afraid. Set up an obstacle course indoors or outdoors that includes a "burrow" to crawl through. Have somebody time you while you speed through your course.

CAMELS (pages 14–15) sometimes carry household items from place to place. Draw small pictures of these items down the left side of a piece of paper: a pillow, a bed, a rug, a table. Then go around your house and count how many of each item you can find. What did you find the most of, pillows or beds? Rugs or tables?

REINDEER (pages 16–17) herds drift north and south as seasonal weather changes make food easier or harder to find. On a calendar, count how many days until the next season arrives. Count how many days until your birthday.

PUFFINS (pages 18–19) can fit as many as twenty fish in their bills. You carry things in your hands. How many items can you carry across a room in one hand without dropping any? Try small, dry things like pebbles or pennies.

AFRICAN ELEPHANTS (pages 20–21) have great memories. Do you? Have a family member put ten random items on a tray. Take a good look. Then have your family member hide the tray. How many of the items can you remember and name?

GIRAFFES (pages 22–23) have too many spots to count. Have a friend pour marbles or pebbles into a glass jar. Get everybody in a group to guess how many pieces there are. Dump them out and count. Who came closest to the correct number?

LEMURS (pages 24–25) are terrific leapers. Using painter's tape, make lines on the floor about 25 inches (64 cm) apart. Count how many space-to-space leaps you can make without touching the lines.

MONARCH BUTTERFLIES (pages 26–27) fly long distances, sometimes as far as hundreds or thousands of miles. What was the longest trip you ever took? How many miles did you travel?

Glossary

Antarctic (ant-AHRK-tik) The area around the South Pole.

Arctic (AHRK-tik) The area around the North Pole.

burrow (BUR-oh) A tunnel or hole in the ground made or used as a home by an animal.

calf (kaf) The young of several large species of animals, such as cows, seals, elephants, giraffes, or whales. The plural is **calves**.

energy (EN-ur-jee) The ability or strength to do things without getting tired.

grooming (GROO-ming) When an animal brushes and cleans another animal's fur.

herd (hurd) A large number of animals that stay together or move together, as in a herd of cattle.

lichen (LYE-kuhn) A flat, spongelike growth on rocks, walls, and trees that consists of algae and fungi growing close together.

migrate (MYE-grate) To move to another area or climate at a particular time of year.

molt (mohlt) To lose old fur, feathers, shell, or skin so that new ones can grow.

shrink (shringk) To make or to become smaller, often as a result of heat, cold, or moisture.

swivel (SWIV-uhl) To turn or rotate around a central point, as in to swivel in your seat.

watering hole (WAW-tur-ing hole) A natural basin of water where animals gather to drink.

Try It! Answers

PAGE 5 7 penguins, less than 10

PAGE 7 8 - 6 = 2 seals

PAGE 9 5 monkeys

PAGE 11 11 + 11 = 22 crabs

PAGE 12 3 + 3 = 6 owls

PAGE 15 12 - 4 = 8 camels

PAGE 17 5 + 10 = 15 reindeer

PAGE 19 10 + 10 + 10 = 30 fish

PAGE 20 12 - 7 = 5 elephants

PAGE 22 6 + 14 = 20 giraffes. The same as 20.

PAGE 25 4 lemurs

Index

Page numbers in **bold** indicate illustrations.

ABOUT THE AUTHOR

Jill Esbaum lives on an Iowa farm and is the author of nearly fifty books for kids. She writes picture books, including *Where'd My Jo Go?*, *We Love Babies!*, *How to Grow a Dinosaur*, and *If a T. Rex Crashes Your Birthday Party*; the Thunder and Cluck series of graphic early readers; and nonfiction books about nature, history, and famous people.